YOUR KNOWLEDGE HAS VALUE

Isata Magbity

Prospect of Bio-fuels in Sierra Leone

GRIN Verlag

Bibliografische Information der Deutschen Nationalbibliothek:

Die Deutsche Bibliothek verzeichnet diese Publikation in der Deutschen National-
bibliografie; detaillierte bibliografische Daten sind im Internet über http://dnb.d-
nb.de/ abrufbar.

Imprint:

Copyright © 2013 GRIN Verlag GmbH
Druck und Bindung: Books on Demand GmbH, Norderstedt Germany
ISBN: 978-3-656-46666-6

This book at GRIN:

http://www.grin.com/en/e-book/230290/prospect-of-bio-fuels-in-sierra-leone

GRIN - Your knowledge has value

Der GRIN Verlag publiziert seit 1998 wissenschaftliche Arbeiten von Studenten, Hochschullehrern und anderen Akademikern als eBook und gedrucktes Buch. Die Verlagswebsite www.grin.com ist die ideale Plattform zur Veröffentlichung von Hausarbeiten, Abschlussarbeiten, wissenschaftlichen Aufsätzen, Dissertationen und Fachbüchern.

Visit us on the internet:

http://www.grin.com/

http://www.facebook.com/grincom

http://www.twitter.com/grin_com

MSc Biotechnology, Bioprocessing & Business

Management

2012/13

Bioproduct Plant Design and Economic Analysis BS937

Prospect of Bio-fuels in Sierra Leone

15/04/2013

Word Count: 1628

Executive Summary

Sierra Leone is a major location for tropical plants such as cassava, palm oil and sugarcane. Its favourable weather conditions, plentiful supply of resources, large arable lands and assets to irrigation assets make it an ideal location for biofuel investment. The potential market for biofuel within the country is extremely small due to the costs of biodiesel and the lack of Government incentives; however its location on the coast of Africa and its vicinity to Western and African countries means offsets this drawback. The most suitable location for an ethanol plant in Sierra Leone is Kambia and for biodiesel production the ideal locations are Pujehun and Kenema. The proposed processing route for the production of cane sugar is fermentation. The recommended process for biodiesel production is an enzymatic transesterification process. This method is adopted to minimize the environmental impact of production. Nevertheless there are still several environmental concerns that arise during the production of biofuels. The most important of these are food security issues. Although the paper concludes that prospects for biofuels are large in Sierra Leone, investors must adopt an emergent strategy to minimise the risks.

Biofuels present a novel opportunity for African countries. They can contribute to energy security, gross domestic product and rural income. They are also important for national export growth. Demand for biofuel is estimated at 170 million litres per annum; however annual demand is projected to reach 230 million by 2020. This essay reviews prospects for biofuels in Sierra Leone. It explains the key issues that arise in the introduction of a biofuel plant in Sierra Leone. These issues are discussed are: resources, policy measures, market opportunities, location and size of plant, processing routes, and finally the environmental and social issues.

Sierra Leone's location at the heart of Western Africa makes it an ideal home for tropical plants such as palm oil and sugar cane. The country's positive agro-climatic condition, a cultivable agriculture area of 4.4 million hectare and its abundant supply of bio-resources means that it can produce biofuels within the quantity required to compete globally (SLIEPA; 2010). At 7°-10° north of the equator, the country has the perfect weather conditions (high annual average rainfall and a mean temperature of 27°C) for growing feedstock such as fresh water algae, sugar cane, palm oil, molasses and cassava. Figure 1 shows the average annual temperature and rainfall. Between January and March, the average rainfall falls below the required amount but this deficit can be overcome by irrigation, which can be provided by any one of the country's 9 river bodies (SLIEPA).

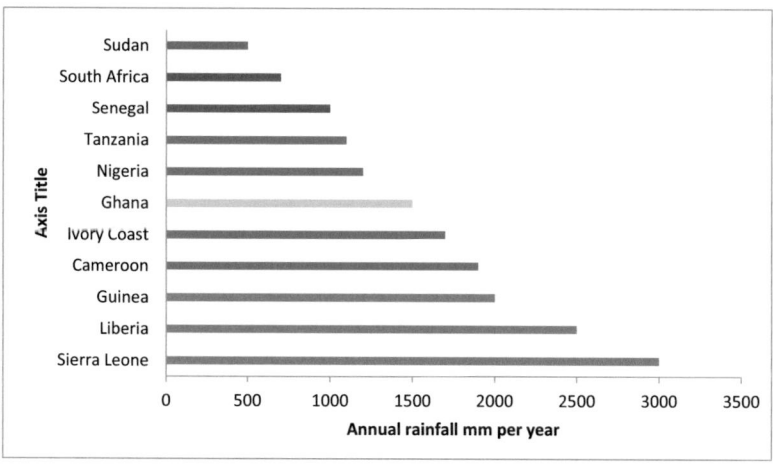

Figure 1: Annual rainfall per year compared to other African countries (SLIEPA; 2010)

Among the resources found in the country, sugarcane is believed to be the lowest cost feedstock with an average cost of $0.35 (Mitchell, 2011). Sugarcane offers several benefits as a feedstock: sugar that can be converted to ethanol; the by-product of the extraction process bagasse can be used to fuel the factory; high sugarcane yields averaging about 85 t/ha; and high ethanol yields of approx. 8000 litres of ethanol per hectare of ethanol. The fact that the country has a lack of high quality sugarcane is a disadvantage however suitable international varieties such as cane Q88 and S13 can also be used (coastal and environmental service; 2009).

Palm oil tree is also suitable for use as a bio-resource for biodiesel production or for use as a straight vegetable oil. A single tree can reach 20 metres. It is the second highest yielding bio-resource in the world with an estimated 3,136 litres of biodiesel per hectare. The drawback of using palm oil as a fuel source is the cost relative to sugarcane ($.20 higher than sugarcane) (Mitchell, 2011). Another limitation of palm oil is the high cloud point of biodiesel sourced from palm oil trees. This means that it can only be used in warmer countries and during the summer seasons in colder countries.

The Government of Sierra Leone has expressed strong support for biofuel investments in the country. State support is demonstrated by the provision of adequate frameworks such as a zero per cent corporation tax law, long lease periods and cheap cost of land (figure 3) and free use of water resources. Labour costs are also competitive at $2 to $5 dollar per day (figure 2). A key motivation for investment is the fact that Sierra Leone also has duty free access to the USA and EU under everything but arms initiative (SLIEPA; 2010). Another investment incentive is the availability of funding from development finance institutions such as the African Development Bank and the Belgian Development company. Swiss company Addax Bioenergy recently received a total of $142 million funding for its sugarcane biofuel project in Sierra Leone (Bio Invest).

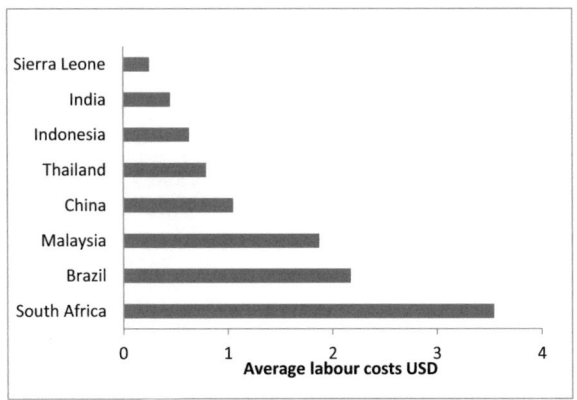

Figure 2: Average labour costs per hour compared to other competing countries (SLIEPA; 2010)

Another driving factor for investment is the potentially large market for biofuel in Sierra Leone. Ethanol can be used to replace charcoal, wood and kerosene. The market size for this in Sierra Leone is great because about 80% of households in Sierra Leone use these products as a primary fuel source (Mitchell; 2011). The market in Africa if bioethanol replaced 10% of the total market of the energy consumption for cooking is roughly 3 billion litres of ethanol (Mitchell, 2011). Second, the by-product of ethanol, bagasse can also be used as fuel in power stations to provide electricity for the factory and neighbouring villages. The expected profits are high because electricity in the country costs about $0.54 per KWh and there is a current deficit of 300 MW (SLEIPA; 2010). The biggest demand for biodiesel is for use as liquid transport to replace gasoline and diesel. The gasoline market is approx. $3 billion and consumption was 134T gallons per year (Algeno Biofuels: 2011). If effective Government policies are introduced, there is a potentially large market for biodiesel and ethanol as fuel enhancers and substitutes for diesel and petrol. Figure 4 shows the projected demand for biofuel from 2010-2020. The supply deficit by 2020 will be about 46 million litres. There is insufficient demand for biofuels in the country and other African regions, however Sierra Leone's coastal location and proximity to Western markets means that companies can still access and compete in international markets (Pinto; 2011).

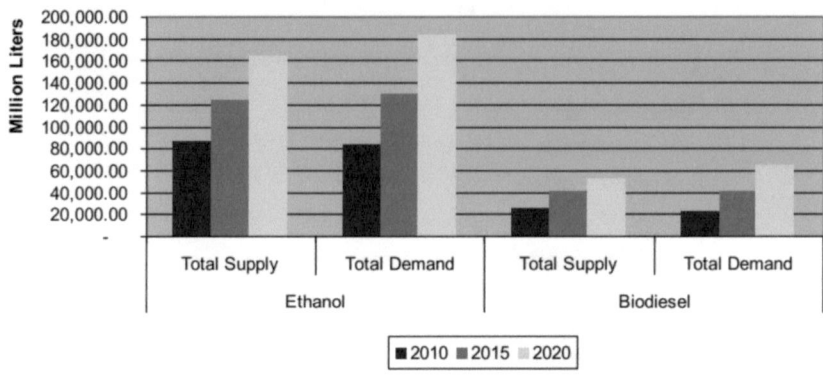

Figure 3: Total biofuels supply and demand (Pinto; 2011)

The plant size and location are two of the most important considerations in determining the feasibility of a bioenergy plant in Sierra Leone. Kambia is the most suitable locations for growing cane (figure 4) because it is in close vicinity to Freetown port (138 km) and it is located near the Kolente River which provides access to irrigation. These areas have an excess amount of land therefore requiring the least amount of resettlements. They also avoid sensitive wildlife areas such as forests and wildlife habitats (SLIEPA). The ideal locations for growing oil palm in Sierra Leone are Kenema and Pujehun because of their proximity to the Mao River (Figure 5) (SLIEPA; 2010).

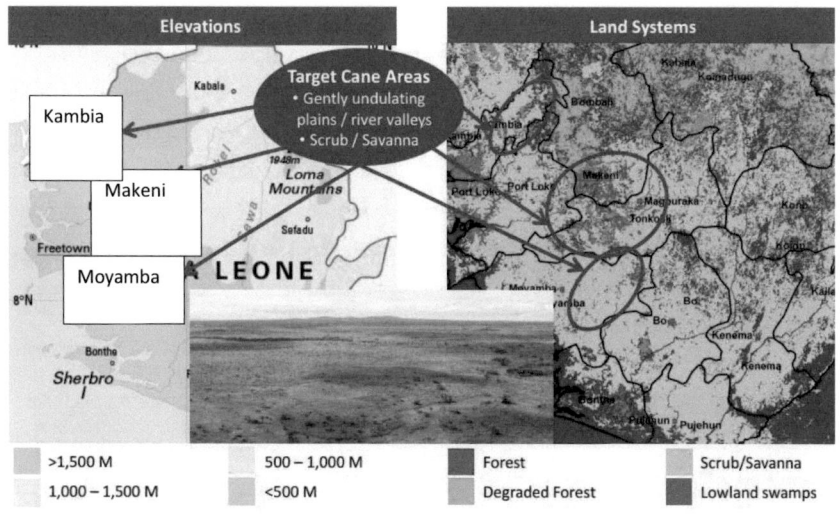

Figure 4: Map of suitable locations in Sierra Leone for an ethanol plant (SLIEPA)

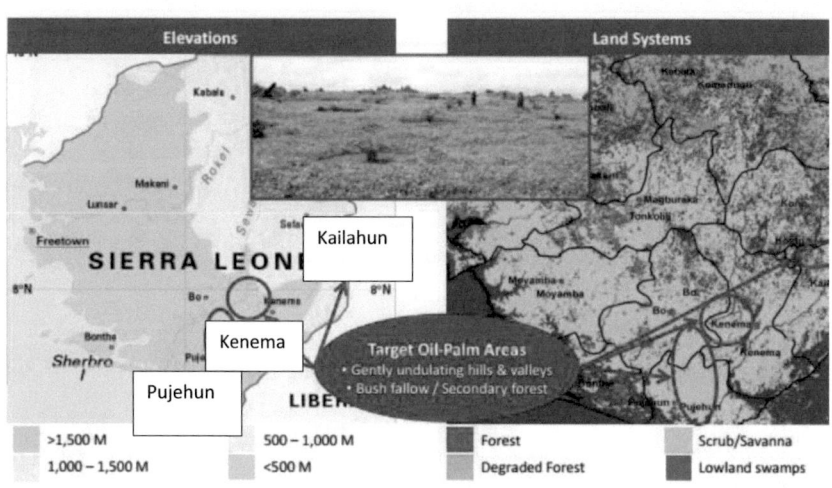

Figure 5: Map of suitable locations in Sierra Leone for a biodiesel plant (SLIEPA; 2010)

Size estimates for the plant are based on specifications for the Addax bioenergy sugarcane to ethanol project. Based on the Addax project, the optimum land size including land for ecological corridors, infrastructure and agriculture is approx. 12000 hectare (African development bank group). A sample plant layout is provided in figure 6. Transportation is a critical factor in the production of biofuels because large quantities of feedstock must be delivered to factory for processing. Suitable roads are required for access to the factory and for transporting fuel and agriculture products and machinery to and from the factory but these will have to be developed privately because of the lack of financial resources to support road developments (Coastal and environmental services; 2009).

Figure 6: Sample plant layout (CES)

The proposed processing route for the fermentation of sugarcane to ethanol is a six step method illustrated below. In phase 1 cane is chopped and shredding to make it easier to use and process. Phase 2 is the extraction step where the sugar juice is removed from cane and bagasse is produced. During clarification the extracted liquid (sucrose) is allowed to stand to sediment all impurities. Yeast is added to the sucrose and placed in a fermenter at 32°C to produce ethanol. 59% ethanol is produced and 41% carbon dioxide. A distiller is used to boil the liquid at 76°C to produce 96% ethanol. The final process is distillation where the ethanol is dehydrated to form 99.7% anhydrous ethanol (Coastal and environmental services; 2009). Figure 6 provides a flow chart of the processing route.

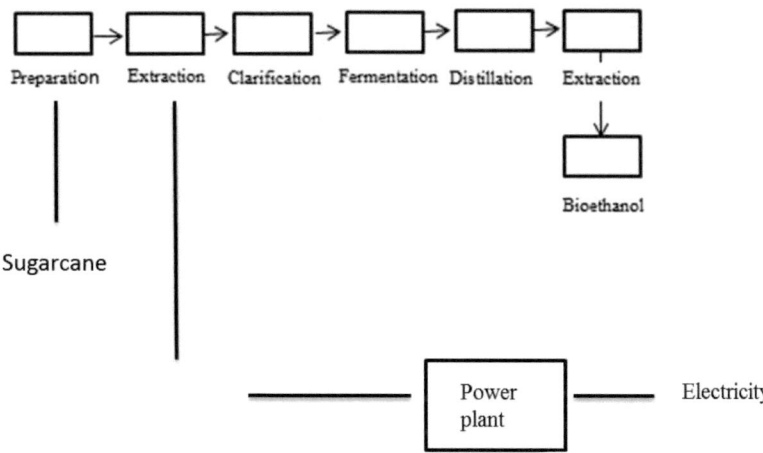

Figure 6: Fermentation of ethanol

The recommended processing route for the transesterification of palm oil to biodiesel is the enzymatic method. This is because the existing chemical method involves excessive energy requirements and produces undesirable by-products. The general reaction of the transferification process is shown below:

$$\text{Vegetable oil} + \text{Methanol} \xrightarrow{\text{Catalyst}} \text{Biodiesel} + \text{Glycerol}$$

The process is illustrated in figure 7. At the initial stages of the process, alcoxy is formed by reacting lipase with methanol. The alcoxy is then reacted with palm oil to produce biodiesel and glycerol. Glycerol settles at the bottom that the biodiesel is removed without disturbing the glycerol sediments. The enzymes used to catalyse the reaction should be immobilised so that they can be reused. The key advantage of this process is its relatively low energy requirements. Disadvantages include the high costs of enzymes compared to chemicals and the inhibitory effects of methanol on the reaction rate (Ranganathan et al; 2008).

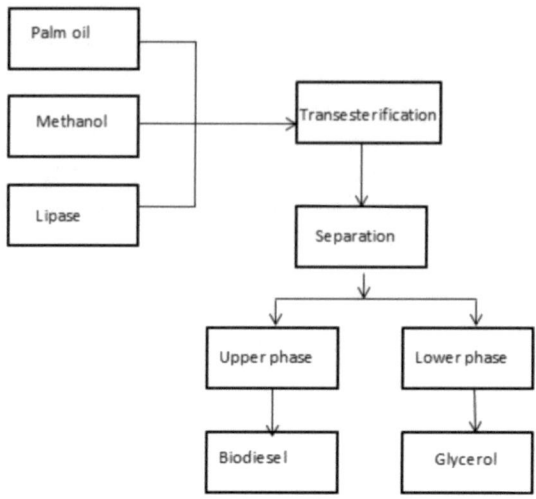

Figure 7: Production of biodiesel using enzymatic catalysis

The potential social and environmental impacts of the project are great. For example cane burning is a health threat because the cane releases harmful gases into the atmosphere such as nitrogen (Mitchell, 2011). Palm oil plantation also has social impacts because it requires a lot of irrigation and the quantity of water required will be drawn from the Mao river. This restricts the community's access to water. Other environmental concerns include issues about land rights, loss of biodiversity, the contamination of the environment and food security. Potential solutions include buying state owned land to minimise the number of resettlements, providing adequate habitats for plants and animals and introducing monitoring programs to effectively manage contamination (African development bank group).With regards to food security issues, sugarcane is not typically consumed as a primary food source. There are however major concerns about the use of palm oil to which is a consumed as a primary food source in many African regions. A possible solution to this is for investors to build development lands so communities can grow their own food.

The above paragraphs review the prospects of biofuels in Sierra Leone. The favourable agro-climatic conditions, supportive state policies, large arable lands, coastal location, proximity to the EU and Americas, duty free access to EU and Americas and an abundant supply of natural resources make Sierra Leone an attractive prospect for biofuel. However the lack of

Government incentives for stimulating demand means that future investors should adopt an emergent strategy that unfolds at each stage of the development. This method would allow the appropriate policy support and regulatory requirements to be developed in phases resulting in fewer risks for investors.

References

African Development Bank Group, Executive Summary of the Environmental, Addax Bioenergy Project: Social and Health Impact Assessment

Algeno Biofuels (2011), Ethanol Market: Demand, [online] Available from: http://www.algenolbiofuels.com/commercialization/ethanol-market [Accessed: 11 April 2013]

Belgian Investment Company for Developing Countries, Addax [online] Available from: http://www.bio-invest.be/en/portfolio/details/93.html?mn=5 [Accessed: 11 April 2013]

Coastal and Environmental Service (2009) Sugar cane to ethanol project Sierra Leone: Project Description, CES: Grahamstown

Mitchell, D. (2011) Biofuels in Africa Opportunities, Prospects, and Challenges, The World Bank: Washington DC

Pinto, M. (2011) Global Biofuels Outlook 2010-2020, Hart Energy [online] Available from: http://www.unece.lsu.edu/biofuels/documents/2013Mar/bf13_04.pdf [Accessed: 8 April 2013]

Ranganathan, S. Lakshmi, S. and Muthukumar, N. (2008) an Overview of Enzymatic Production of Biodiesel, Bioresource Technology, Vol. 99, Issue 10, pp. 3975-3981[online] Available from: http://www.sciencedirect.com/science/article/pii/S0960852407003999 [Accessed: 3 April 2013]

Sierra Leone Investment and Export Promotion Agency (2010) Sierra Leone Investment Outreach Campaign: Opportunities for Investors in the Oil Palm Sector [online] Available from: http://www.investsierraleone.biz/download/SL_OilPalm_Investment_Opp.pdf [Accessed: 3 April 2013]

Sierra Leone Investment and Export Promotion Agency, Sierra Leone Investment Outreach Campaign: Opportunities for Investors in the Sugar Sector [online] Available from: http://www.investsierraleone.biz/download/SL_Sugar_Investment_Opportunities.pdf [Accessed: 3 April 2013]